ABOVE: *Modern machinery can produce nails at an amazing speed.*
FRONT COVER: *Cut nails being made at the Tremont Nail Company, Wareham, Massachusetts, on a machine built in the 1840s. The worker feeds in a plate of steel, turning it over after each nail has been guillotined to even out the taper of the nails.*

NAILMAKING

Hugh Bodey

Shire Publications Ltd

CONTENTS

Introduction 3
Roman nailmaking 5
The middle ages 7
Slitting mills 11
The domestic industry 14
Factory nailmaking 21

Copyright © Hugh Bodey, 1983. First published 1983. Shire Album 87. ISBN 0 85263 606 7. All rights reserved. No part of this publication may be reproduced or transmitted in any form or by any means, electronic or mechanical, including photocopy, recording, or any information storage and retrieval system, without permission in writing from the publishers, Shire Publications Ltd, Cromwell House, Church Street, Princes Risborough, Aylesbury, Bucks, HP17 9AJ, UK.

Set in 10 on 9 point Times roman and printed in Great Britain by C. I. Thomas & Sons (Haverfordwest) Ltd, Press Buildings, Merlins Bridge, Haverfordwest.

For the many students, especially in Yorkshire and Devon, whose enthusiasm and curiosity have led me to this and other interests.

This nailmaker's workshop has been reconstructed at the Avoncroft Museum of Buildings, Bromsgrove.

FRENCH PATENT FORGED NAILS,
ST. MARCEAU WORKS.
ONLY GOLD MEDAL Awarded for Excellence of Quality, PARIS EXHIBITION, 1878.

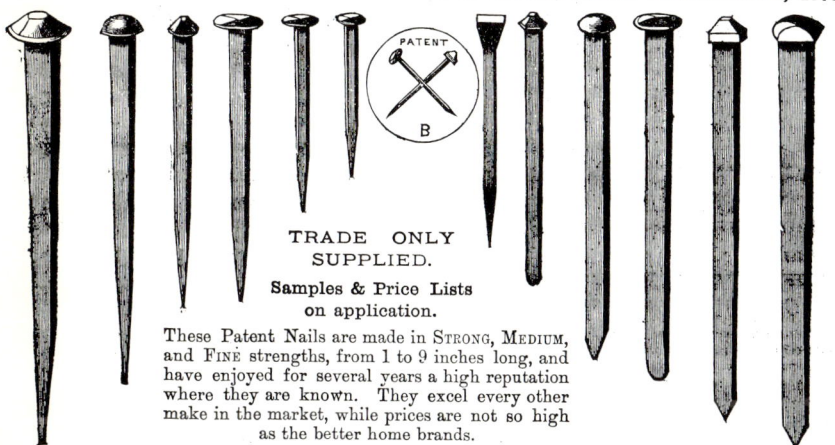

TRADE ONLY SUPPLIED.

Samples & Price Lists on application.

These Patent Nails are made in STRONG, MEDIUM, and FINE strengths, from 1 to 9 inches long, and have enjoyed for several years a high reputation where they are known. They excel every other make in the market, while prices are not so high as the better home brands.

H. & F. BONTEN, 116 Queen Victoria Street, LONDON.
GENERAL AGENTS TO THE ST. MARCEAU NAIL WORKS, FOR HOME AND EXPORT.

An advertisement from 'The Ironmonger', 1885.

INTRODUCTION

Nailmakers produced a wide variety of nails and fixing devices, though not screws, bolts or rivets. Many of their goods, of course, were common nails, pointed at one end and with a head at the other, and hammered into place. Yet not all nails fit this standard description. Some nails have no flat head; they are designed to sink below the surface of the wood into which they are being hammered so that they will not show, and either have very small heads like panel pins or no heads at all. Some nails have domed heads and are put in as much for decoration as to fasten. And there are many nails that are heavily ridged in some way so that they will give better grip.

Further problems arise when trying to draw the limits between nails and other fixing devices. Screws are clearly quite different, and so are rivets for a rivet only works when the tail of the shank is reshaped so that it has two heads. Nails do not alter their shape to achieve their purpose. Spikes, though, are really large nails. Some shelf supports, for instance, were designed to be hammered straight into walls — they were ornamental spikes. Sometimes they were made by blacksmiths, at others by nailmakers. Most blacksmiths made nails at times, especially in areas distant from the main nailmaking centres. This book concentrates on the full-time nail and spike makers.

Some kinds of nails are fast disappearing from daily use. Carpet tacks have given way to gripper rods as wooden floors have been replaced by concrete. As the use of plastic coatings has increased, nails have been replaced by adhesives. Labour savings on building sites include interlocking tiles in place of nailed slates, and factory-made roof frames fastened with clips. New shapes of nails are being introduced, for use with conventional or pneumatic hammers.

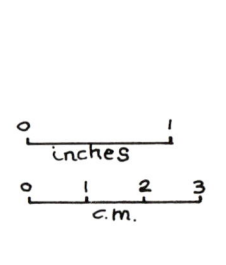

RIGHT: *A selection of Roman nails recovered from excavations. Some have been reclaimed, to judge from their shapes.*

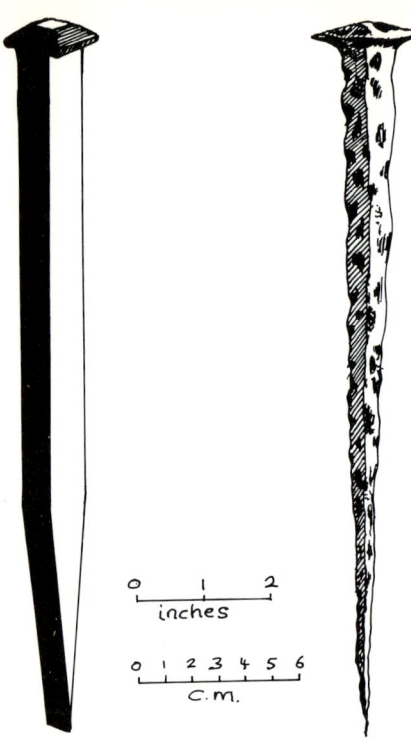

ABOVE: *A Roman nail from the Inchtuthil excavation (right). The head seems to have been made by hitting the iron with a die. To the left is a 6 inch (150 mm) hand-made spike from a boat built in 1812.*

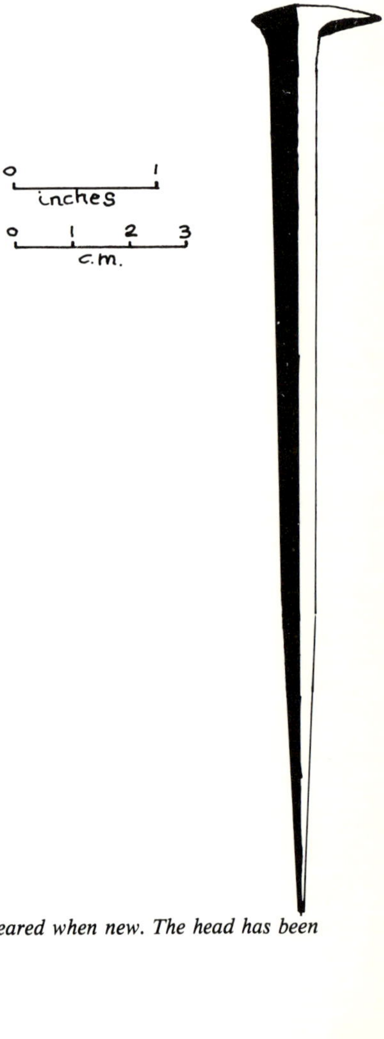

RIGHT: *A Roman nail drawn as it would have appeared when new. The head has been formed by bending the iron and hammering it.*

ROMAN NAILMAKING

The ironworking Celts, from whom the centuries before the Roman invasion take the name 'iron age', do not seem to have used their iron for nails. Iron was still scarce, so much so that it was used as a form of currency. Bronze nails had been used in earlier times to build boats but the widespread use of nails in Britain began with the Roman troops. It was not that nails suddenly became cheaper, rather that the military budget allowed the legions to have whatever they wanted. The troops pressed on into the frontier regions, building base forts as they went. The barrack blocks were prefabricated and came from legionary stores. Such buildings could be assembled very quickly. When the fort became redundant because the frontier had advanced further or could not be held, buildings were dismantled. The nails could be extracted with claw hammers and returned to the stores for reuse — they were not cheap enough to throw away. However, when they abandoned the camp at Inchtuthil near Aberdeen, 12 tons of nails were left in a heap.

Each nail seems to have been made individually. A piece of iron was heated in a charcoal furnace and hammered to shape on an anvil. The cross-section of the nail was therefore square, the easiest shape to make by these methods. The iron was reheated as often as necessary, and the shank was made to taper along its entire length – to fit it for use as a square peg in a round hole. The head was fashioned last, and it is possible that this was done by placing a mould on the end of the shank and striking it hard – this would account for some of the lop-sided heads that can be seen.

Many other types of nails were made at this time, by the same methods. One of the most common was the small hobnail that covered the soles of soldiers' sandals. These were almost a weapon – as a wedge of troops pressed forward into the enemy, they deliberately trampled on wounded people lying on the ground to crush the life out of them. Other tacks were made to hold roof tiles in place, to fix hinges on doors, to fasten wood to make buckets and baskets – the uses in Roman times have many parallels in the nineteenth century.

This medieval illustration of the building of Noah's Ark shows the techniques of house construction used in the middle ages. On the roof wooden trenails are being used on the right, iron nails on the left.

A medieval blacksmith beating out a bloom to make a bar of wrought iron. It had to be constantly reheated in the charcoal furnace to keep it workable.

THE MIDDLE AGES

The supply of nails declined after the Roman occupation ended, although their production and use continued. Footwear, particularly for farmers, still relied heavily on nails to fasten layers of leather together. The richly ornamental hinges on church doors, fixed with nails, are an indication that iron was scarce, to be used for special purposes only, but the range of products of the nailer began to extend beyond just common nails. These were still made, but there is evidence of a growing output by more specialised makers. One example is the spikes used in naval battles during the Hundred Years War, before guns altered the methods of warfare. Before that time, enemy ships roped themselves together and the fighting was done by marines. The object was to drive the rival crew overboard in order to capture a valuable boat: injury was directed against people, not the ship. One way of spreading confusion was to throw powdered lime from the crow's nest and let the wind carry it over the other ship. While the crew were rubbing their eyes, four-legged spikes were thrown on to the deck. These had sharp points, so that they lodged in the woodwork — there was always one spike sticking up and seamen worked barefoot.

Because the history of nailmaking has never attracted much interest in the past, little evidence about the industry in the middle ages has survived. But it seems that by the fifteenth century, perhaps before, there were specialist nailmakers in the West Midlands, and possibly in other areas too, since nails were heavy in relation to their value and local production must have been preferable to the high cost of transport. The methods of making nails seem to have changed in order to produce more. Instead of shaping each nail from a piece of iron, the iron was first hammered into a sheet of the thickness of the finished nails and then cut into rods with hand shears. Iron was smelted in bloomeries at this time, and a bloom of iron from the furnace weighed 15 or 20 kilograms (30-40 pounds). This would have been a manageable size to flatten into a sheet. The process was only used for the smaller sizes — tacks and pins — because coldshear iron is brittle, hard to cut and useless for large spikes. It was small nails that were needed in the greatest quantities, though, and the changed methods must have speeded production of these considerably. After the rods were cut, the required length was cut off, heated in a charcoal fire and hammered to a point. The final stage, shaping the head, was done by gripping it in a vice so that little of the iron could spread sideways and striking it with a heavy hammer. There is evidence for iron nails being made in the west midlands at this time, and the other iron-producing regions – Kent and Sussex, the Forest of Dean in Gloucestershire, south Yorkshire and Lancashire, Ayrshire and so on – may also have been making nails.

Brass nails were also being made at this time. Heavy ones were used in ship construction because of their greater resistance to corrosion, and decorative tacks were also being used in, for example, some of the earliest chairs. Brass can be shaped at lower temperatures, which was a help. The large Cistercian abbey at Tintern on the Wye produced brass wire, and it is likely that this was the raw material for some nailmakers. These, however, were a different and much smaller group of men than the nailers who made iron nails.

A scarf joint in a late medieval house, held in place by trenails.

ABOVE: *Wooden pegs were still used in the nineteenth century to hold heavy stone slates on the roofs of weavers' cottages. These can be seen at the Colne Valley Museum, Golcar, West Yorkshire.*

BELOW: *A fine oak chest made about 1500. The dome-head nails held both the straps and hinges and added to the decorative effect.*

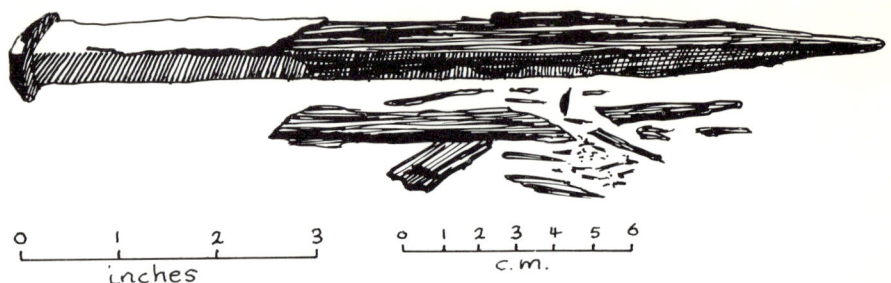

ABOVE: *Wrought iron is very strong but hard wear and exposure to the air lead to corrosion. This follows a distinctive pattern, almost like the layers of an onion.*

BELOW: *A spike on a late nineteenth-century clay barge shows the same kind of wear and fatigue.*

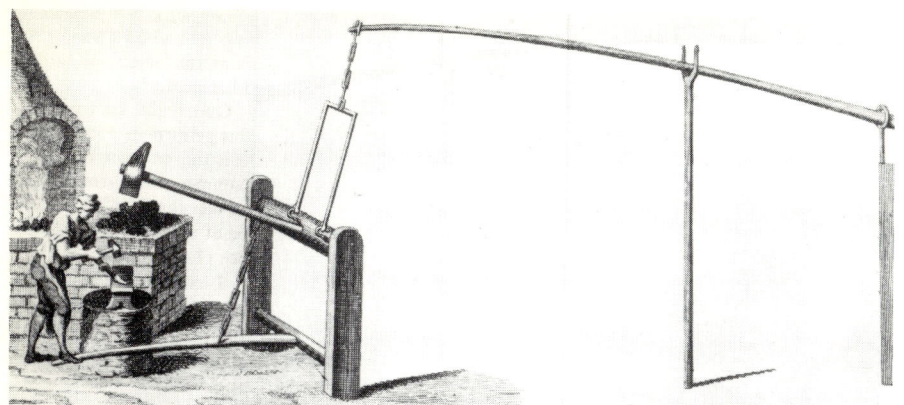
The oliver, a treadle-operated heavy hammer used in nailshops for shaping the heads of the nails.

SLITTING MILLS

The demand for nails increased markedly in the sixteenth century. Construction of timber-framed houses mostly required wooden trenails but the laying of floor-boards called for iron nails in increasing quantities as floors were inserted in previously open halls towards the end of the century. The population was rising again after two centuries of rapid decline and people were becoming more prosperous, and this inevitably increased demand. Henry VIII and Elizabeth I both expanded the navy – indeed they created it – and it used a great many nails, and many other shipyards were busy as Britain suddenly embarked on international trade. The pressures may well have led to higher prices, since the output of nails could not be increased rapidly because of the laborious process of making rods. Birmingham was one of the few towns where an absence of guild and manorial restrictions allowed enterprising craftsmen to expand their activities and increase their output. A tireless traveller, Leland, passed through Birmingham in the 1540s and noticed that there were 'many smithes in the towne that use to make knives and all maner of cuttynge tooles, and many lorimars that make byts, and a great many naylors. So that a great parte of the towne is mayntayned by smithes', who obtained their 'yren out of Staffordshire and Warwikeshire and see cole out of Staffordshire'. The use of local coal in the blacksmiths' forges enabled them to overcome the difficulties caused by the scarcity of charcoal and its rising price. Indeed access to such suitable coal gave the West Midlands the lead in nail production, which it was to regret later. Birmingham was the largest centre for iron goods but they were made in many surrounding towns and villages also. The son of a successful nailer in Wednesbury, for example, became Lord Paget, who expanded iron production on Cannock Chase in the 1540s.

The preparation of iron rods was accelerated by an idea imported from Liege, the slitting mill. The first one in Britain was set up by Godfrey Box in Kent in 1590. The slitting mill used water power to hammer (later to roll) blocks of heated iron into sheets and then to shear them into rods. The gradual spread of slitting mills to other ironworking areas transformed nailmaking. All sizes of nails could now be cut from prepared rods, greatly reducing the time and effort needed for making nails. This was reflected in reduced prices, which further stimulated demand. From being a product with some scarcity value, nails quickly became a principal consumer of iron and a valuable addition to exports.

While the growth of production seems to have taken place in all the nailmaking areas, if only to keep pace with demand, it was the West Midlands where expansion was most rapid and which provided the bulk of the export trade. It is also the area about which most is known, partly because of later enquiries into the area in the nineteenth century, when nailmaking became synonymous with poverty. (For that reason, the area is used in much of this book to illustrate what happened: there were many other areas making nails, such as the villages around St Helens in Merseyside.) The fame of the West Midlands spread far, even to the West Indies, where someone commented in 1657 that 'Nailes of all sorts with hooks, hinges and cramps of iron . . . are to be had at Birmingham in Staffordshire much cheaper than in London'. Both towns were markets for nails and similar goods; production was carried on elsewhere. The evidence suggests that Dudley was the centre of nailmaking in the sixteenth and seventeenth centuries. For example, twenty out of the fifty-four Dudley men who appeared before the quarter sessions between 1591 and 1643 were nailers; the more obvious occupation, farming, came second with seventeen men. Dudley, though, was not a dominant centre, more a leader among equals. Other nailmaking towns and villages in the region at this time were Sedgley, Rowley Regis, Wednesbury, Coseley, Netherton and Newcastle-under-Lyme. West Bromwich had joined this group by the middle of the seventeenth century — nineteen of the twenty-five bridegrooms married there in 1656 and 1657 were nailers. Some nailmaking had been done there in previous centuries but it now became a major centre.

By the middle of the seventeenth century nailmakers were becoming poorer. Many nailers in the sixteenth century had been comfortably off, if not wealthy. The background of Lord Paget has been mentioned already, and a number of nailers from the West Midlands villages were able to put down sureties of £10 each in the 1580s and 1590s — a substantial sum of money. By the end of the seventeenth century, though, nailing had become a low paid trade. There were several reasons for this. The main one was the large number of slitting mills that had been set up on the river Stour in the Midlands, as well as in other parts of Britain. Much of what in the middle ages was called the 'mistery' of nailmaking lay in blending iron to make it suitable for nails, and in reducing large blocks to a size and shape that could be fashioned into the required end products. Nearly all the skilled part of the business was now handled in the slitting mill. Richard Foley set up the first Midlands slitting mill at Stourbridge in 1628 and was soon followed by others. They used much local iron but also bought in iron of different kinds from outside the area, including cinders which were brought up the Severn from Roman iron-smelting sites in the Forest of Dean. The slitting mill masters then produced the varying qualities of iron suitable for different purposes, all in finished rods of the necessary cross-section. These were being rolled by the end of the seventeenth century, just as girders are rolled in a twentieth-century steel mill. All the nailer had to do was reduce the rod to short lengths and shape them into nails. There was no longer much money to be made at that.

Another consequence of dividing the process of nailmaking into work done in two separate places was that the capital requirements of the nailers were now much less. Until the end of the sixteenth century nailers had to have an extensive forge and finery in order to handle the blooms of iron that they bought. (Some had their own bloomeries as well, so that they could smelt their own iron – the Pagets, for example.) They also had to have anvils and small tools for shaping the nails and held substantial stocks of nails for sale. All this cost money, which limited the number of people able or willing to enter the craft, and craft guilds added further restrictions. The relatively small number of nailers in any one place could easily be regulated, if only by mutual agreement, so that price fixing and limitations on output could be practised. All this was changing by the middle of the seventeenth century and continued to change into the eighteenth. The nailer now needed only a small forge to heat the end of the rod, and tongs and hammers. The whole equipment cost little and could fit in an outhouse. No capital was tied up in raw materials or completed stocks. Consequently larger numbers of men took up nailing. The

ABOVE: *This leather-covered travelling trunk made by Edward Smith about 1730 is a fine example of the decorative use of brass nails.*

RIGHT: *Dome-headed brass nails add elegance to this 1780s chair as well as fastening the leather seat to the mahogany frame.*

demand for nails, both in Britain and overseas, continued to rise, maintaining high prices, so the trade seemed to be one worth joining. Also the success of nailers in previous generations had given the trade a status above that of some others. Within a few decades, however, that status and the high prices had declined. The much larger number of men in the trade could not be controlled, the influence of the old guilds was over and the scattered geographical spread of the nailers added to the difficulty. The ease of entry to the trade attracted both those who were ironworkers and men who knew nothing about it. They could normally obtain a few rods from the mills on credit, and renting a corner of a nailmaker's workshop cost only pence a week. Entry to the trade could hardly have been easier. The result was that the supply of nails came to exceed demand and the income of the nailers fell to little more than subsistence level, compelling them to work long hours in order to stay alive. This in turn attracted less desirable characters into the organisation of the industry, the hated middlemen called foggers. The decline of the nailer was a gradual process extending into the nineteenth century. Demand, as for any item, varied from year to year, and people so closely connected to farming were accustomed to taking good and bad years together. In addition, the demand for nails was still growing as the colonies increased, particularly those in America which were major customers. One third of the 20,000 tons of iron produced in Britain in 1700 was made into nails – there seemed to be plenty of work for all.

THE DOMESTIC INDUSTRY

By the end of the eighteenth century, nailmaking provided work for tens of thousands of people in the Black Country alone. When Arthur Young travelled from Birmingham to West Bromwich in 1776, he noted 'for five or six miles it was one continuous village of nailers'. There were now four slitting mills on the Stour, two on the Tame and two on the Churnet. An estimated 35-40,000 nailers converted 10,000 tons of rods into nails and spikes in 1799, the bulk of them near the Stour valley in places like Kingswinford, Wordsley, Brockmore, Brettel Lane, Gornal, Sedgley and Rowley, while there were centres in the Tame valley at Darlaston, Oldbury, West Bromwich and Coseley. As James Keir noted, nailers did not have to live in these towns to carry on their trade: 'As this manufacture required a very simple apparatus of a small hearth, bellows, anvil and hammer, it is executed at the workman's own house, to each of which houses a small nailing shop is annexed, where the man and his wife and children can work without going home; and thus an existence is given to an uncommon multitude of small houses and cottages, scattered all over the country, and to a great degree of population, independently of towns.'

A number of other changes had come about by the end of the eighteenth century. It was now common for nailmaking to be combined with farming. It was not normally farming by day and nailmaking at night so much as full-time farming when the season demanded, at haymaking and harvest, and nails for the rest of the time. Much, though, depended on the demand for nails and therefore the price the nailers got per thousand. The other change is that women and children were now participating in nailmaking – scarcely possible in the sixteenth century and rare in the seventeenth. This may sound cosy, like any description of domestic industry, which gives the impression that the whole family could choose to work when they would, and that they could enjoy each other's company while making their living. The reality was otherwise. A whole family working full-time might have £1 to £1.50 a week after working from 4 a.m. to 10 p.m. five days of the week (Tuesday to Saturday, little work being done on Sunday or Monday). However, that would only be while demand was brisk; at other times the same hours might yield only 70p, well below subsistence level. Demand was so brisk towards the close of the French wars that nailmakers worked 72 hours a week

The reconstructed nailshop at the Abbey House Museum, Kirkstall, Leeds (key below): 1, a late nineteenth-century flypress, used to shape heads more speedily than by using the oliver (6); 2, bellows for the forge: the handle is long enough to be worked while the nailmaker holds metal in the fire with his other hand, and the air passes to the forge through the pipe; 3, stone tub for water to cool metal quickly and so harden it; 4, the forge where iron rods were heated in a charcoal fire (coke later in the nineteenth century); 5, long-handled tongs for holding hot iron, and shorter ones for iron being worked on the anvil (7); 6, the oliver, a heavy hammer held up by a springy pole and brought down by a treadle: the face of the oliver could be fitted with different dies so as to be able to make several shapes of nails, and the nail was held fast to be struck; 7, the anvil.

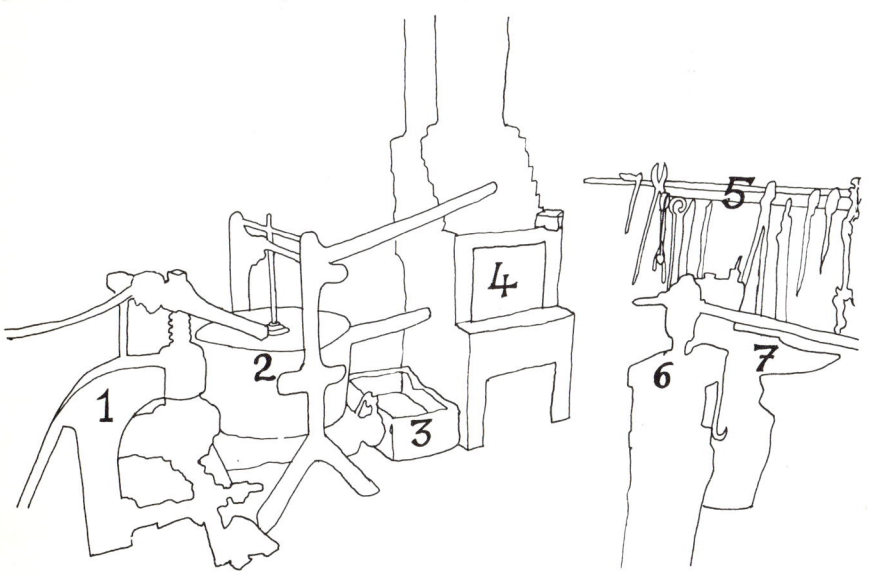

A bundle of rods and a box of finished nails in the nailmaker's shop in the Abbey House Museum, Kirkstall, Leeds.

over 347 days in 1810, 80 hours a week over 347 days in 1811, and 81 hours a week over 353 days in 1814. Though demand was high, wage rates had not risen in line with wartime inflation – nailers were working harder but not prospering.

The industry was controlled by nailmasters, often the owners of slitting mills, such as Thomas Green whose mills slit 12,000 tons a year. Other masters were merchants of the finished nails. They put work out to hundreds of nailers each. Nailers were thus dependent on the masters for work from week to week – they collected a bundle of rods weighing half a hundredweight (25 kg) on Monday and took back the nails on Saturday. If there was no work on offer from that master for the next week they had to travel around other masters to find some, or do none and have no income that week. The masters either sold their nails direct to wholesalers and exporters or found other outlets: Thomas Green sold some of his to ironmongers and the rest through the big fair that had grown up in Stourbridge. Increasingly in the nineteenth century foggers took a larger part in the running of the industry. They would often hand out work when the masters' warehouses were full but it was on their own terms – foggers were notorious for giving out shortweight iron, for using rigged scales when the nails were brought back, for allowing little for wastage and for paying in tokens that could be exchanged for goods only at a shop controlled by the fogger. By these means they were able to undercut the regular merchants but it was sweated labour for the nailers – enforced slavery to earn a living for lack of any other local occupation. Not that the nailers were entirely blameless: many would claim more for wastage than was the case and sell the extra nails themselves.

The collapse of other forms of domestic industry, particularly spinning, added to the numbers of women and children who worked in nailing in the early decades of the nineteenth century, and this tended to push prices down further. Working conditions grew worse, as described in the 1843 Report to Parliament:

'The best forges are little brick shops of about 15 feet by 12 feet (4.5 m by 3.7 m), in which seven or eight individuals constantly work together, with no ventilation except the door and two slits, a loop-hole in

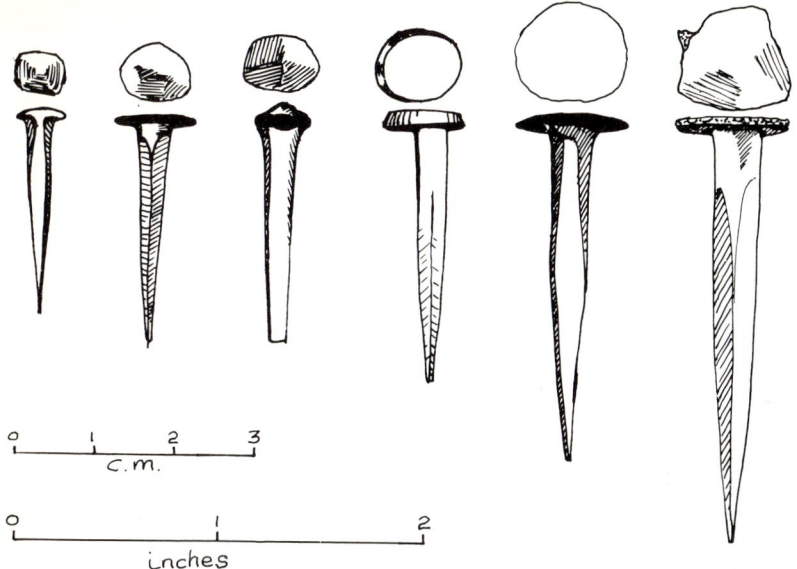

ABOVE: *The common nails used in making clogs.*

BELOW: *Howard Bamforth, a trained clogger, at work in the reconstructed clogmaker's workshop at the Colne Valley Museum, Golcar, West Yorkshire.*

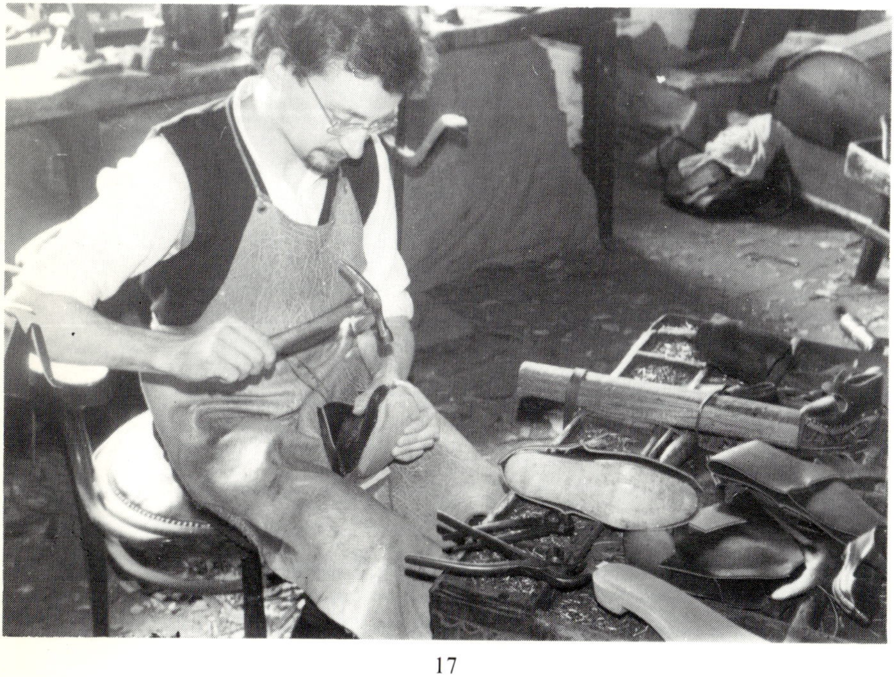

the wall; but the majority of these workplaces are very much smaller, filthily dirty, and on looking in upon one of them when the fire is not lighted, presents the appearance of a dilapidated coalhole.

'In the dirty den there are commonly at work a man and his wife and daughter, with a boy or girl hired by the year. Sometimes the wife carries on the forge with the aid of her children. The filthiness of the ground, the half-ragged, half-naked, unwashed persons at work, and the hot smoke, ashes, water and clouds of dust, are really dreadful.'

Parliament did nothing to improve conditions – the problem was more complex than anything in their experience of factory regulation so far. The problem was not a new one, unlike the cotton mills, and so the nailers drifted lower. The decline was made faster by the sudden drops in demand, first when supplies to America were stopped during the War of Independence and later when the Americans began to make their own nails. The end of the wars with France in 1815 caused the navy to cancel orders. To make matters worse, the first successful attempts to make nails by machine started at this time also. Though the making of horse nails and of large nails and spikes still called for the strength of a man, women increasingly made the tacks, hobnails and other small wares, while their men sought work elsewhere. The anonymous contributor to *Knight's Pictorial Gallery* in 1858 gives the best description of Black Country nailmaking, though he ignores the meagre returns for those who relied solely on their nailmaking:

'Whoever has occasion to pass along the public roads between Birmingham and the adjacent towns of Dudley, Walsall, Wolverhampton, &c, will be pretty sure to see indications of wrought nail making. Here and there an open door will afford a glance into a rude kind of smithy or shop where three or four persons are hammering away, and where a smithy fire affords the means of heating the iron. Along the road, too, may be seen persons carrying bundles of iron rods, which they have purchased of the iron merchants, and are about to convert into nails at their own dwellings. The rods are made of iron which has been rolled into sheets of the requisite thickness, and then cut up by slitting rollers into pieces having the requisite width for nails. The working up of these rods into nails is an operation in which all the members of a family frequently assist. In the first place, one end of a rod is placed in the forge-fire, and heated by the aid of two or three blasts from small bellows. The nailor takes it out of the fire and rests it on his anvil, which is a small cube of steel imbedded in a mass of iron. With a few dexterous blows from his hammer he quickly fashions the end of the rod into the required shape for a nail; and then cuts off the portion thus prepared. Another heating and another hammering produce a second nail; and so he goes on until the whole length of the rod is exhausted. By certain simple tools which he employs, the nailor is enabled to give any desired shape to the nail, and to fashion one end of it into the form of a head. The celerity with which all this is effected almost surpasses belief. There was one occasion on which a nailor undertook to make seventeen thousand large nails in a week, for two weeks together; a feat which he successfully accomplished . . . the above quantity is allowed to be as much as

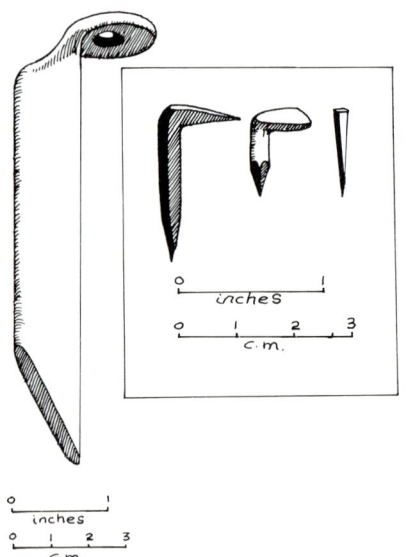

A variety of hand-made spikes: (left to right) holdfast, tenterhook, cooper's clasp, glazier's sprig.

Among the many crafts requiring nails of various types and in considerable quantity was that of the wheelwright. The engraving on the right, taken from 'The Book of English Trades and Useful Arts' (1827), shows a wheelwright nailing iron strakes to a wheel to form a tyre. The photograph above, taken in the wheelwright's shop at the Ashley Countryside Collection, Wembworthy, near Chulmleigh, Devon, shows a less specialised use for nails, as pegs or hooks for holding various items of the craftsman's equipment. Collectors of old nails will find overhead beams like this a good source of unusual specimens.

three ordinary men can perform without difficulty.'

Women were expected to produce as many as men, though of the smaller sizes. Children were allowed three months to learn the trade and were then expected to make a thousand nails a day. They were often cruelly treated, by their parents rather than by anyone else, and only the coming of compulsory education after 1870 rescued them.

The domestic nailmaking industry in the West Midlands operated in a similar fashion to other cottage industries in the area, such as chainmaking and odd-work (the production of hand-made objects such as latches and hinges). Indeed the same families often engaged in all three activities, turning more to chainmaking and odd-work as nailmaking declined. This social pattern was not paralleled in the other iron-producing regions.

Nailers tried striking in the 1850s to raise wage rates but no strikes achieved permanent success, until the Franco-Prussian War (1870-1) cut off imports of Belgian nails and there was a short-lived boom. A nailer in Sedgley, who had earned as much as £1.50 a week in the 1830s, was averaging 45 pence in 1880. Census figures of Staffordshire nailers indicate the general decline: 1861, 9,016; 1871, 7,424; 1881, 5,583; 1891, 2,726; 1901, 1,057. The handmaking of nails lingered on into the twentieth century, associated with the making and repairing of shoes by hand, for cobblers preferred hand-made nails for the soles. By the 1920s the centuries-old craft had gone.

Cobblers' nails: (above, from left) heel tip and nail, flat stud, triple hob, two rivets, lasting tack; (below from left) two cut nails, single hob.

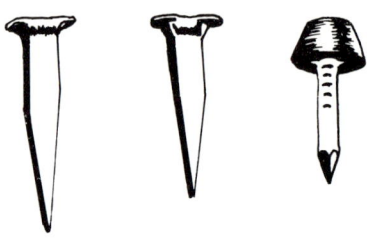

A nail factory in the 1930s. Overhead shafting drives these machines, which make small nails.

FACTORY NAILMAKING

CUT NAILS

Production of nails in Britain did not cease when the old hand trade died out. The first nail-cutting machine was set up in 1811, and factory-made nails were being produced in such numbers by 1830 that they were adding to the hardships already described. The first machine-cut nails were made from plate iron. Strips of iron were rolled and slit so that they were the thickness of the finished nail, and as wide as the nails would be long. The strips were heated and held against a powerful guillotine worked by overhead shafting moved by water or steam power. The trade was carried on in workshops from the start, for it could not be grafted on to domestic working. The machine cut off a tapered shape, which was the cut nail. The machine operator turned the iron sheet over, and the guillotine sliced off the next nail. By the 1840s machines had been developed which cut the nails so as to have a small head, and in 1866 a machine that could cut four rows of nails at a time from a strip came on to the machine tools market. The price of nails fell sharply with the use of these methods, the more so when imports from Belgium and later Germany began. Harrods sold half-inch (13 mm) tacks for less than a penny a thousand in 1895.

Cut nails varied from the very small sprigs used to hold window panes in position while the putty hardened to 6 inch (150 mm) long ones used to fix floor joists in place. The headless nails sank into timber flush with the surface – one use was to nail floorboards down. The correct way

Heavy machinery in a nail factory in the 1930s. Wire was drawn in, straightened and shaped, and the finished nails fell into small bins on the floor.

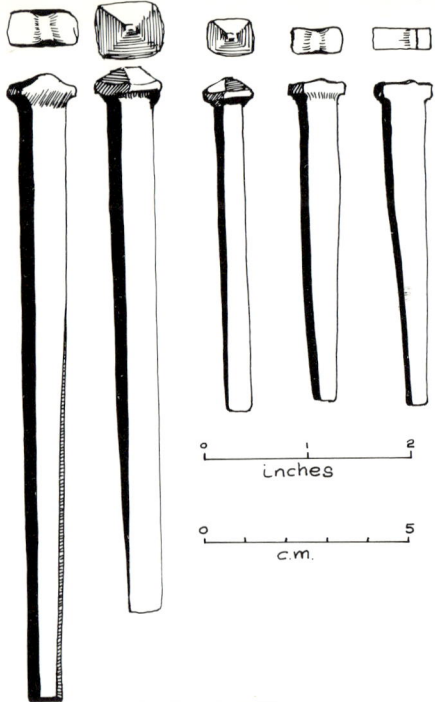

ABOVE: *Cut nails: (from left) T-head nail – note the chamfer effect of the punch, two rose-headed fine nails, two clasp nails – used to hold casings in place.*

to do this was to nail them through the edge of the board into the joist, so that no nail head was visible. This method made it very difficult to lift the floor again. Cut nails can be found in most older houses and indeed are still made on both sides of the Atlantic. Earlier cut nails can be distinguished from the later stamped ones if the nail is still in good condition. The older ones have the edges sliced off at right angles to the face; later ones have the edges drawn down slightly by the punch – almost like a chamfered edge.

WIRE NAILS

While the cut nail was still growing in popularity, the first machines to make nails from steel wire were being tried. This has since become the principal way of making nails because of the high speed and continuous runs that can be achieved. Such machines reduce nailmaking to two actions which follow each other at lightning speed. The first is striking the end of the wire with a die to form the head of the nail, which can come in many shapes. When the die has swung back, the nail is cut from the wire at its correct length, and the action of cutting produces the point. A works in Cardiff has two hundred such machines, which collectively make 1,400 nails per second. The machines of a Lennoxtown firm make smaller nails at the rate of six hundred a minute on each machine.

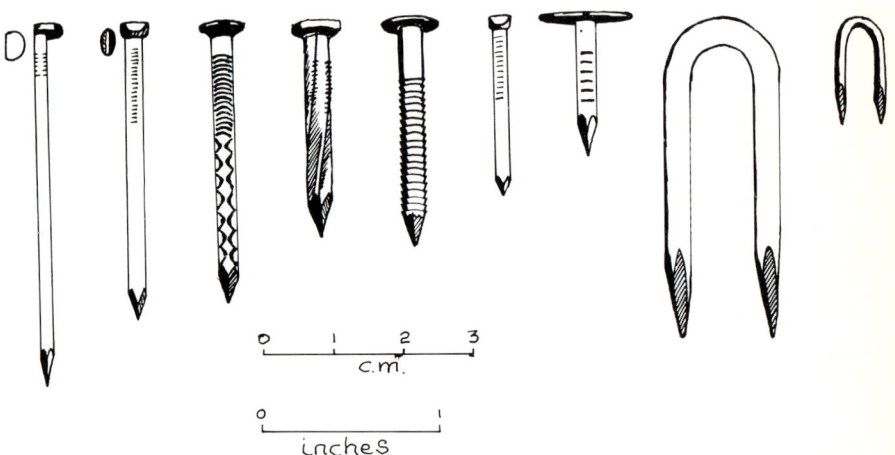

Bright steel wire nails: (from left) round plain head (for pneumatic hammer), oval brad, plasterboard nail, square twist, large-head annular ring, oval, extra large head clout, two staples.

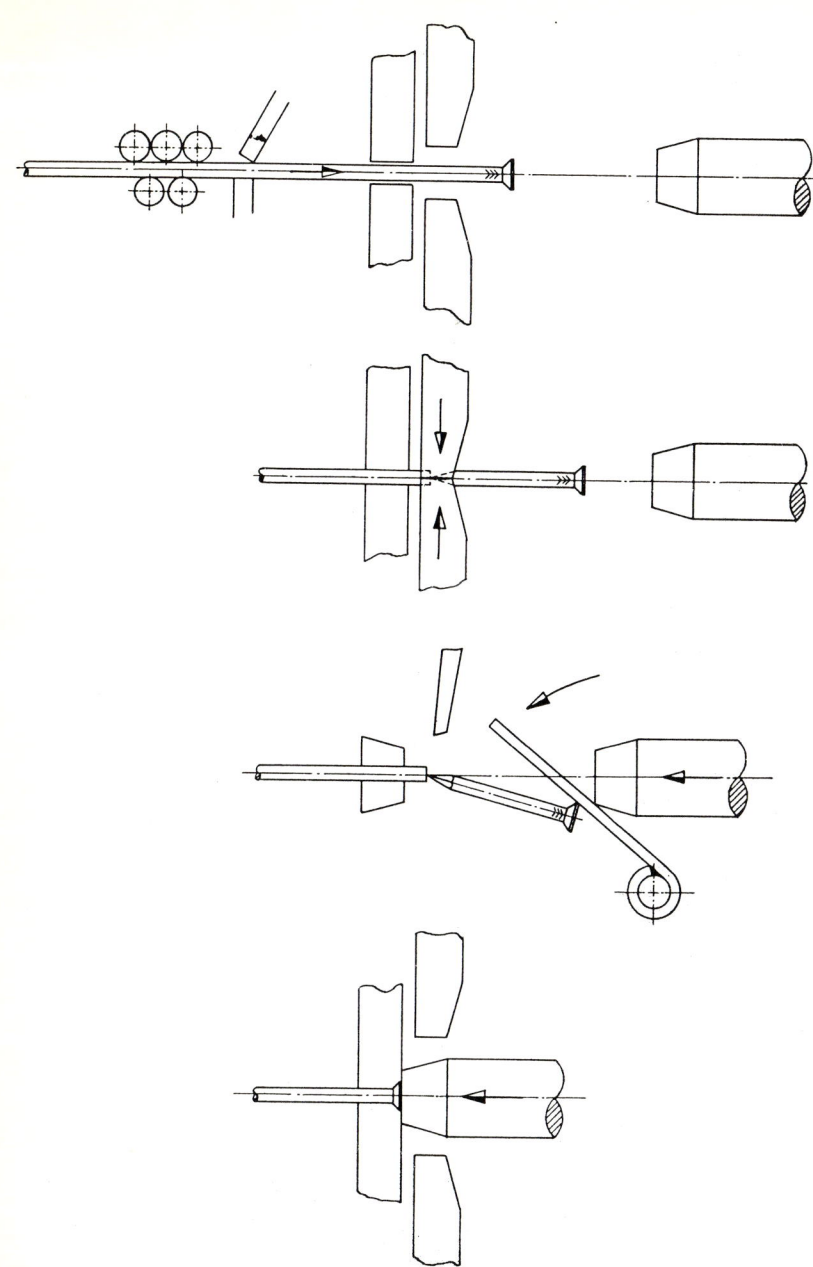

The sequence of operations in a high-speed nailmaking machine: (from the top) pulling in the correct length of wire; cutting the wire and so forming the point; ejecting the finished nail; forming the next head.

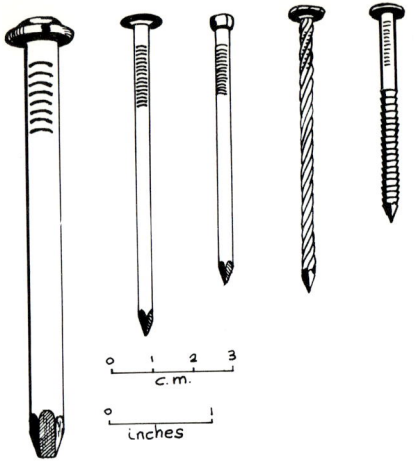

ABOVE: *Bright steel wire nails: (from left) pipe nail, clout, lost head, helical threaded, annular ringed.*

Most nails are made from round wire but some are made from square wire. A certain amount of twist is put into these as they are drawn between stages one and two, so that the nails will grip better in use, such as the springheads used to hold corrugated asbestos on roofs. After the nails have been ejected from the machines, they are cleaned in a bath of hot caustic soda to remove grease and small scraps of metal. Some are then polished by being rotated in drums of hot sawdust. Others are coated by being galvanised or sherardised to make them resistant to corrosion. For sherardising, the nails are heated to 300 C (570 F) in a powder of zinc oxide and zinc dust. The slightly rougher surface produced by both processes also gives the nails a better grip. The finished nails are packed by magnets so that they lie parallel in cartons for transport to the customer.

COLLECTING NAILS

Though some older shapes have gone, others have taken their place. Collecting nails is a relaxing hobby which has the advantage that collections do not take up much room. Anyone interested in forming a collection of modern nails is recommended to buy a 5 pound box of assorted nails — at least twenty kinds can be expected, often far more. Older nails can be recovered from do-it-yourself jobs and demolition sites (always ask first though), the beams of sheds and garages where they have been used for hooks and pegs, and from old tool chests. Some districts are fortunate enough to have old-established ironmongers, whose stock in trade is a useful source. Watch out when they retire or move shop.

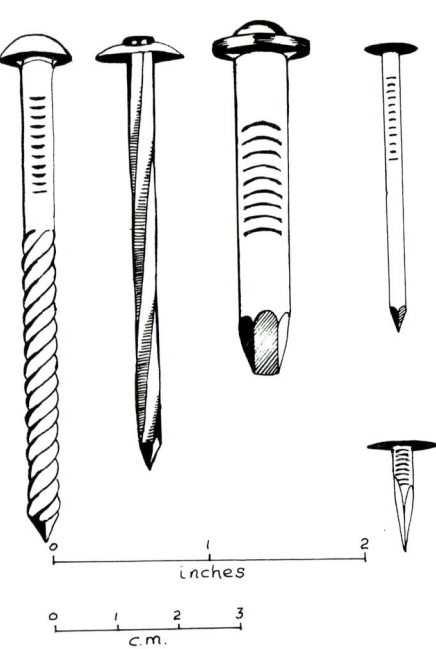

LEFT: *Sherardised wire nails: (from left) helical thread, spring head, pipe nail, lath nail, extra large head clout (bottom right).*

Machinery installed in 1947 at a Warrington nailmaking factory. Individual electric motors replaced the shafting and the bulk bins travel underneath the gangways.

ABOVE: *Much of the high-speed nailmaking machinery used in Britain today is made in Germany. This compact machine shapes between 220 and 900 nails per minute, depending on the size being made.*

BELOW: *Machinery making OBO masonry nails. The nails are made in the machine on the left, hardened at a temperature of 840 C (1,540 F) in the centre, and then coated with zinc in the automatic barrel plant on the right.*

Nails used to be sold in sacks, weighed and sewn on a machine like the one shown here. Nowadays, however, they are packed in cartons.

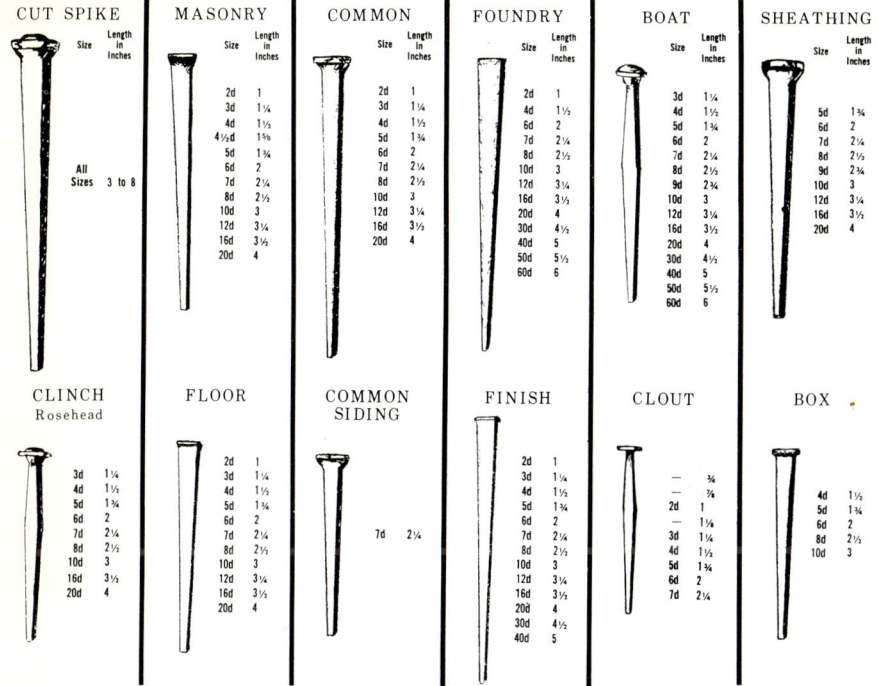

The Tremont Nail Company of Wareham, Massachusetts, established in 1819, is one of the oldest businesses in the United States and the world's oldest nail manufacturer still in production. Some of the kinds of cut nails the company makes are shown above. The company's main building (left) was constructed of timber by ships' carpenters and the walls are faced with unpainted shingles. The cupola on the roof housed a bell that was sounded to call the workers to the factory. Old-style nailmaking machinery at the factory is pictured on the front cover of this book. Visitors are welcome at the factory.

FURTHER READING

Black Country nailmaking is well served:
Allen, G. *Industrial Development of Birmingham and the Black Country, 1860-1926.* Cass, 1966.
British Association. *Birmingham and its Regional Setting.* 1950.
Court, W. H. B. *The Rise of Midland Industries, 1600-1838.* Oxford, 1938.
Timmins, S. *The Resources, Products and Industrial History of the Birmingham and Midlands Hardware District.* 1865.

Other areas are poorly served, and a search through local histories and such useful but time-consuming publications as 'notes and queries' or local newspapers is necessary. Also helpful is D. Bythell *The Sweated Trades.*

For American nails but useful to British readers also is Lee H. Nelson *Nail Chronology as an Aid to Dating Old Buildings.* This well illustrated leaflet can be obtained from the American Association for State and Local History, 1400 Eighth Avenue South, Nashville, Tennessee 37203.

Further information and illustrations of nails may be found in other Shire Albums, notably *Old Horseshoes, The Victorian Ironmonger, The Village Blacksmith* and *The Village Wheelwright and Carpenter.*

PLACES TO VISIT

Abbey House Museum, Kirkstall, Leeds. Telephone: Leeds (0532) 462632. The only complete nailmaker's shop in Britain has an oliver, sets of tools and samples of products.
Avoncroft Museum of Buildings, Stoke Heath, Bromsgrove, Worcestershire. Telephone: Bromsgrove (0527) 31363. A short row of workshops has been re-erected here.
Black Country Museum, Tipton Road, Dudley, West Midlands. Telephone: 021-557 9643/4. Some nailmaker's tools.
Cliffe Castle, Spring Gardens Lane, Keighley, West Yorkshire. Telephone: Keighley (05352) 64184.
Hunterian Museum, University of Glasgow, Glasgow. Telephone: 041-339 8855 (extension 285). Roman nails.
Rockbourne Roman Villa, Rockbourne, Fordingbridge, Hampshire. Telephone: Rockbourne (072 53) 445. Roman nails.

Examples of Roman nails can also be seen at other site museums and in a few specialist museums.

Readers in the United States can visit the Tremont Nail Company, 21 Elm Street, Wareham, Massachusetts. There are no such opportunities in Britain.

ACKNOWLEDGEMENTS

The author is grateful to Humphrey Boyd for the care he took with the drawings, the Ashley Countryside Collection, Wembworthy, Devon, for facilities to photograph, and the following for permission to reproduce illustrations: Avoncroft Museum of Buildings, page 2; British Library, pages 6, 7, 11; Castle Nails, page 1; Mr G. E. Holloway, pages 8, 9 (top), 17 (bottom); Hunterian Museum, University of Glasgow, page 5; Douglas Kane Ltd, page 28 (bottom); Leeds City Museums, pages 15 (top), 16; Museum of English Rural Life, University of Reading, page 19 (bottom); Rylands-Whitecross Ltd, pages 21, 24, 27, 29; Sotheby's, pages 9 (bottom), 13 (both); Tremont Nail Company, front cover and page 30; Wafios Maschinenfabrik, Wagner Ficker and Schmid, Reutlingen, pages 23, 28 (top). Other photographs are by the author.

The publishers acknowledge with gratitude the advice of Mrs Jennifer Costigan of Avoncroft Museum of Buildings in the preparation of this book.